Spring Patterns
ADULTS COLORING BOOK

Karen Ciaccia

This book belongs to :

_____
_____
_____
_____

Copyright 2022 © Karen Ciaccia - Visual Medicine

All rights reserved. No part of this publication may be reproduced, distributed or, transmitted in any form or by any means, including recording or other electronic or mechanical methods.
https://www.amazon.com/author/karenciaccia

COLOR TEST PAGE

www.ingramcontent.com/pod-product-compliance
Lightning Source LLC
Chambersburg PA
CBHW080459220526
45465CB00006B/2316